IMAGES
of America

THE
BROOKLYN
NAVY YARD

THE USS QUINNEBAUG. The steam sloop of war USS *Quinnebaug* is shown here on what appears to be laundry day. The *Quinnebaug* was launched at the Brooklyn Navy Yard in 1866. (USNI.)

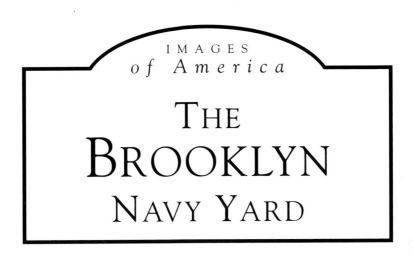

IMAGES
of America

THE BROOKLYN NAVY YARD

Thomas F. Berner

ARCADIA
PUBLISHING

Copyright © 1999 by Thomas F. Berner
ISBN 978-0-7385-5695-6

Published by Arcadia Publishing
Charleston SC, Chicago IL, Portsmouth NH, San Francisco CA

Printed in the United States of America

Library of Congress Catalog Card Number: 2008922467

For all general information contact Arcadia Publishing at:
Telephone 843-853-2070
Fax 843-853-0044
E-mail sales@arcadiapublishing.com
For customer service and orders:
Toll-Free 1-888-313-2665

Visit us on the Internet at www.arcadiapublishing.com

To my wonderful parents and wife.

BATTLE FLEET. The U.S. battle fleet is gathered off Annapolis, Maryland, in 1913. Although most of the ships are not identified in the navy's archives, several ships were almost certainly built at the Brooklyn Navy Yard. The first three ships are the USS *Rhode Island*, the USS *New Jersey*, and the USS *Georgia*. (NHC.)

Contents

Acknowledgments ... 6

Introduction ... 7

1. The Age of Sail: 1801–1861 ... 9

    Dateline: February 7, 1801 ... 12

2. The Age of Transition: 1861–1904 ... 29

    In the Shoals of History: Footnotes to the Past ... 48

3. The Age of the Battleship: 1904–1941 ... 57

4. The Age of Total War: 1941–1945 ... 83

    Dateline: January 31, 1941 ... 88

5. The End of an Age: 1945–1966 ... 107

    Dateline: September 25, 1966 ... 115

Epilogue ... 119

Appendix: Ships Constructed at the Brooklyn Navy Yard ... 127

ACKNOWLEDGMENTS

The maps on pages 122 and 124 are courtesy of the U.S. Geological Survey (USGS) and the satellite photographs on pages 123 and 125 are courtesy of Aerial Images Inc. and Sovinformsputnik via TerraServer URL (*www.terraserver.com*). All of the other items in this book come from sources attributed in the captions: the Brooklyn Public Library's Brooklyn Collection (BPL), the Library of Congress (LC), the Museum of the City of New York (MOCNY), the National Archives (NA), the Naval Historical Center (NHC), the New-York Historical Society (NYHS), the Smithsonian Institution (SI) (LC, NA, NHC, and SI photographs were received through the Naval Historical Foundation (NHF) and the U.S. Naval Institute (USNI). The staffs of each of these institutions (including Elizabeth Ellis and Marguerite Lavin of the MOCNY, David Manning of NHF, Borinquen Gallo, Holly Hinman, Laird Ogden, and Nicole Wells of the NYHS, and Stacey Niemann of USNI) were so very helpful, that I can claim only textual errors as exclusively my own. I would especially like to thank the "three Js" of the Brooklyn Public Library: Julie Moffat, Joy Holland, and Judy Walsh, who are gracious, cheerful, and knowledgeable—traits that are not normally associated with the Borough of Brooklyn, except by those who know better. Because BNY did not just build ships but also served as a base for the fleet, many of the ships pictured in this book were not built at the yard. A complete list of ships built at BNY is set forth in the appendix.

BATTLESHIP DIVISION TWO. On June 7, 1954, two battleships built at the Brooklyn Navy Yard, the USS *Iowa* and the USS *Missouri*, are joined by the USS *New Jersey* and the USS *Wisconsin*. (NA.)

INTRODUCTION

There are places in this world that ache with history, places that have had such a profound effect on who we are and how we got this way, that the past has never left them. History lingers in the corners of such places like the sweet smell of narcissus on a windless day. To enter such a place is to be smitten with the poignant sense that one is about to physically reenter the past.

One such place can be found along a slightly shabby stretch of New York City's waterfront on the edge of Wallabout Bay at what was once the Brooklyn Navy Yard. Its official name was originally the New York Navy Yard, then the U.S. Navy Yard—New York, and finally the New York Naval Shipyard; but Brooklynites have always called it by its unofficial name, the Brooklyn Navy Yard (referred to hereafter as BNY). Now a city-owned industrial park for private industry, BNY was one of the U.S. Navy's most important shipyards for 165 years. It was a participant in every military buildup except for the earliest, in the 1790s, and the latest, in the 1980s.

BNY played a much larger role in naval history than its cramped 219 acres (plus 72 acres of water) would indicate. Any list of the five most famous warships in American history will probably contain four ships that were commissioned here: the *Monitor*, the world's first modern warship; the *Maine*, whose sinking caused the Spanish-American War; the *Arizona*, whose sinking launched America into World War II; and the *Missouri*, on whose deck World War II ended. The first transatlantic cable was strung from the stern of a ship built here. The U.S. Naval Academy was founded because of a mutiny on a BNY-built ship. BNY was a leader in the navy's transition from sail to steam-driven ships.

Navy yards were America's first large-scale industrial complexes, combining a largely civilian work force with military command, all supported by government funding. Besides providing port facilities and provisions for the fleet, and constructing and repairing naval vessels, navy yards were responsible for the manufacture of many of the necessities for outfitting ships. BNY made many items, ranging from flags and pillowcases to torpedo compressors and turbine blades, in the course of its career. Right up to its closing, BNY manufactured sails, although by 1966, the sails were intended only for the Naval Academy's training ship.

Navy yards occupy a peculiar niche in American economic history. Historians of a leftist bent argue that these government-owned installations put the lie to the myth of an America founded on free enterprise and point to a government role in the growth of American economic power. Conservative historians find these yards to be early examples of government waste, where ships took many years to be built, launched, and commissioned in order to provide pork-barrel politicians with a perpetual source of no-show jobs for campaign supporters. Any boost to the economy, they argue, came about because their low productivity made them easy to compete against. There is, in fact, some truth to both of these claims, but there is no denying that a worthy by-product was the creation of a navy second to none.

One

THE AGE OF SAIL:
1801–1861

Although the U.S. Navy acquired BNY on February 23, 1801, the yard was inactive for several years. No construction was started until 1805, when the first six buildings were built, and BNY did not receive its first commandant until June 1, 1806. It did not launch its first ship until 1820, although during the War of 1812, more than 100 ships were fitted out at BNY. BNY-built ships participated in the invasion and blockade of Mexico during the Mexican War. BNY was on the cutting edge of the technology of the time: the introduction of steam as a means of propulsion. One of the world's first military think tanks was created at the yard when the Naval Lyceum was established in 1833. In the years before the Civil War, BNY launched 21 major ships: one ship of the line, three frigates (one of them powered by steam), nine sloops of war, two schooners, two brigs, three paddle wheel steamers, and one revenue cutter. With the secession of the southern states in 1861 and the loss to the Union of the navy yards located in the South, BNY assumed even greater importance to the fate of the nation.

THE PRISON HULK JERSEY DURING THE REVOLUTIONARY WAR. More than 12,000 patriots died of disease or starvation in the *Jersey* and the 16 other prison hulks, decrepit former sailing ships that were crammed with American prisoners of war. (NYHS.)

QUARTERS A, THE COMMANDANT'S RESIDENCE. Quarters A was one of the first buildings at the Brooklyn Navy Yard. Whether or not the architect of the White House actually designed this residence, it contains a room remarkably similar in shape and size to the Oval Office. Quarters A is now a registered landmark. (BPL.)

THE STEAM BATTERY FULTON. Also known as the *Demologos*, the *Fulton* was the first steamship built for any navy. It was not designed to be seagoing but was built to defend New York Harbor. Built in New York, but not at the Brooklyn Navy Yard, it was BNY's receiving ship from 1815 to 1829. A receiving ship was a retired vessel converted into barracks for new recruits during their training. (USNI.)

THE USS OHIO, FIRST SHIP BUILT AT BNY. The 2,757-ton ship of the line USS *Ohio* was the first ship built at the Brooklyn Navy Yard. The keel was laid in 1817 and the ship was launched in 1820, a fairly short construction period for that era, although it took nearly 20 years before the ship entered the fleet. Although BNY records show the *Ohio* to have carried 74 guns, this print claims that it carried 104. Either way, it was one of the largest ships in the fleet, the battleship of its era. (USNI.)

THE LAUNCHING OF THE OHIO ON MAY 30, 1820. The method of launching—sliding stern-first down an inclined shipway—remained the standard way to launch ships until the 1950s. (NYHS.)

Dateline: February 7, 1801

The East River slices Manhattan Island from Long Island. It is not really a river at all but a tidal estuary, extending from Long Island Sound to New York Harbor, a distance of less than 10 miles. At one point, in a Brooklyn neighborhood known as Fort Greene between the neighborhoods of Williamsburg and Brooklyn Heights, the river makes a wide swing to the west, carving out a crescent-shaped bay measuring about a half mile by a half mile, which is bounded by two hills and into which flow two creeks.

The Canarsie, the Native-American group that occupied most of present-day Brooklyn, called this spot *Rennegachonk*, meaning "sandy beach," or *Mahrenhanreck*, meaning "bay." The Dutch called it *Waal-boght*, from the word "Walloon," because the first settlers in the region were Belgian Huguenots. When the English seized New Amsterdam from the Dutch in 1664, *Waal-boght* was anglicized into Wallabout Bay.

On June 16, 1637, Joris Jansen de Rapelie, a Huguenot, purchased 335 acres of land from the Canarsie tribe. After his death, his wife, Catelya Trico de Rapelie—sometimes called the Mother of Brooklyn for her role in settling the city—continued to farm the land. The de Rapelie family retained the land until 1781, a period of time almost as long as the period the navy occupied it. At some time during their tenure, a toll bridge was established near what was to become the southwest corner of BNY.

The military history of the site began during the Revolutionary War. Gen. George Washington anchored his left flank along Wallabout Creek during the Battle of Long Island on August 27, 1776. With Washington's retreat from New York, the British used Wallabout Bay to station 17 "prison hulks," decrepit former sailing ships crammed with American prisoners of war.

After the war, brothers John, Samuel, and Treadwell Jackson founded a shipyard on the site of the future BNY, and the first vessel, a merchant ship called the *Canton*, was built in 1798. Shortly afterward the frigate *Adams* was built for the U.S. Navy, the first government vessel built there (the *Adams* was burnt by its captain, Charles Morris, on September 3, 1814, at Hampden, Maine, to avoid its capture by the British).

A naval buildup was part of the platform of the Federalist Party in the 1800 presidential election, while malign neglect was the military doctrine of Thomas Jefferson's Democratic-Republicans. When Jefferson won the election, the lame-duck Federalists under John Adams pushed to acquire land for naval bases before they left office on March 4, 1801. Accordingly, on February 7, 1801, John Jackson sold 41.93 acres of waterfront land to Frances Child, as agent of the United States, for $40,000. Sixteen days later, Child sold the land to the navy for $5. (This does not represent altruism on Child's part but was the nominal consideration cited in the documentation transferring ownership of the property from the federal government to the navy.) On May 18, 1801, the city of Brooklyn sold the navy all waterfront rights to the site for $1.

Because Jefferson had opposed a naval buildup, the yard was inactive for several years. Even such an isolated country as the United States, however, could not avoid the fallout from the Napoleonic Wars. Reluctantly, the Democratic-Republicans began to think about defense. The first six buildings at BNY were constructed in 1805, and on June 1, 1806, BNY received its first commandant, Lt. Jonathan Thorn. He commanded the yard until July 13, 1807 (he was killed by Native Americans in Oregon three years later). One of the first buildings, the commandant's residence, known as Quarters A, is believed to have been designed by Charles Bulfinch, the architect of the White House.

THE USS SAVANNAH. The keel of the 44-gun frigate USS *Savannah* was laid in 1820, but the ship was not launched until May 5, 1842. The 22-year construction period indicates that the ship was a lucrative source of featherbedding jobs for politicians' campaign supporters. The *Savannah* was the flagship of Commo. John D. Sloat's fleet, which assisted in the seizure of California during the Mexican War. (USNI.)

THE USS VINCENNES. The 18-gun, 700-ton sloop of war USS *Vincennes* was begun in 1825 and launched on April 27, 1826. The *Vincennes* was damaged and run aground during the Civil War in its battle with the Confederate ironclad CSS *McRae* (known as the "Iron Turtle") on October 12, 1861. It was refloated and served throughout the war. (USNI.)

THE USS PEACOCK. The 18-gun second-class sloop of war USS *Peacock* was launched on September 30, 1828. It was used on the U.S. Exploring Expedition, led by Lt. Charles Wilkes, from 1838 to 1842. It is shown here in Antarctica. (USNI.)

BNY IN 1831. The large barn-like structures were construction sheds known as ship-houses. This is a view of the yard from the northeast side of Wallabout Bay. The 1830s saw a major expansion of the functions of BNY. (NYHS.)

THE NAVAL LYCEUM. The Naval Lyceum was founded at the yard in 1833. One of the first outlets for sharing technological and strategic theories and developments, it published the *Naval Magazine* in 1836 and 1837, the first naval publication in America. One frequent contributor, James Fenimore Cooper, argued the case for a powerful navy. The Naval Lyceum closed in 1877, and its functions were taken up by the U.S. Naval Institute. (NYHS.)

THE MARINE HOSPITAL, FOUNDED AT BNY IN 1834. Throughout the 19th century, additional parcels of land were acquired or sold off as needs required. By 1900, the navy had received over $1.4 million more for selling off portions of the land than it had spent for the entire site. (NYHS.)

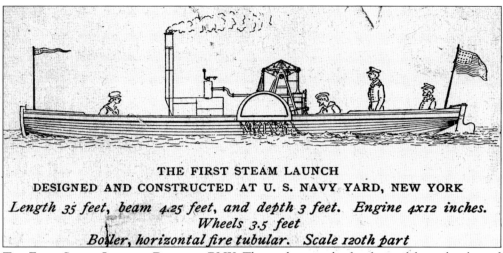

THE FIRST STEAM LAUNCH BUILT AT BNY. The yard was at the forefront of the technological changes that revolutionized shipping in the 19th century. (MOCNY.)

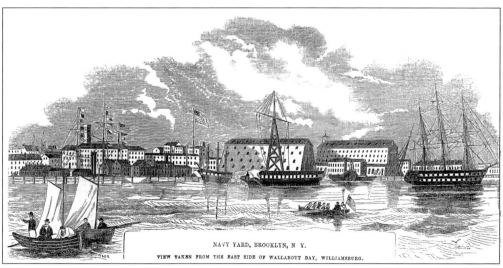

VIEW FROM BNY, C. EARLY 1837. In the center of this drawing of the long pier leading to the Marine Hospital are the covered bows of the frigates USS *Constitution*, in front, and USS *Brandywine*, whose ship's clock later graced the commandant's office until BNY closed in 1966. The ships USS *Fredonia*, left, and USS *North Carolina*, right, are beside the ship-houses. At the time of this picture, the USS *Fulton 2nd* was under construction in the ship-house to the right, and the USS *Sabine* was under construction in the ship-house to the left. (Author's collection.)

THE USS *FULTON* 2ND, FIRST STEAMSHIP BUILT AT BNY. The nine-gun paddle wheel steamer USS *Fulton 2nd* was the first steamship built at BNY. Begun in 1835, it was launched on May 18, 1837. In size and performance, it represented a major technological development over earlier steamships. (USNI.)

THE USS *DECATUR*. The 16-gun sloop of war USS *Decatur* was begun in 1838 and launched on April 9, 1839. Ships of the line were the largest ships in the fleet during the age of sail, followed in descending order (among other categories) by frigates, sloops, and brigs. Each had a function befitting its size. (USNI.)

VIEW OF BNY, 1840. This 1840 view taken near the Brooklyn-Queens border shows BNY in the center background. The bucolic nature of Brooklyn and Queens had not disappeared after more than 200 years of settlement. (MOCNY.)

THE TOMB OF THE MARTYRS. The Tomb of the Martyrs was erected on the edge of the yard in the early 19th century to commemorate those who died in the prison hulks during the Revolutionary War. It was replaced in the late 19th century by the graceful obelisk seen today in nearby Fort Greene Park. (MOCNY.)

TOMB OF THE MARTYRS, BROOKLYN, LONG ISLAND.

CAPT. MATTHEW CALBRAITH PERRY. Capt. Matthew Calbraith Perry, known as the Father of the Steam Navy for his efforts to modernize the fleet, was commandant of BNY from June 12, 1841, to July 15, 1843. The brother of famous naval hero Oliver Hazard Perry, Matthew Perry led the fleet that opened Japan to foreign trade in 1854, and inadvertently taught the Japanese the value of modern weaponry and naval power. (NYHS.)

THE USS SOMERS. The 10-gun brig USS *Somers* was launched on April 16, 1842. It began its career as a school ship, an experimental program to educate youngsters to be naval officers. Something went wrong. A mutiny occurred on November 26, 1842, and three mutineers were hung, two of whom can be seen just below the flag. One of the mutineers, Midn. Philip Spencer, was the son of the secretary of war. Since this proved to many that teenagers were too young to be trained at sea, the U.S. Naval Academy was founded. The *Somers* was lost in a storm during blockade duty in the Mexican War. (USNI.)

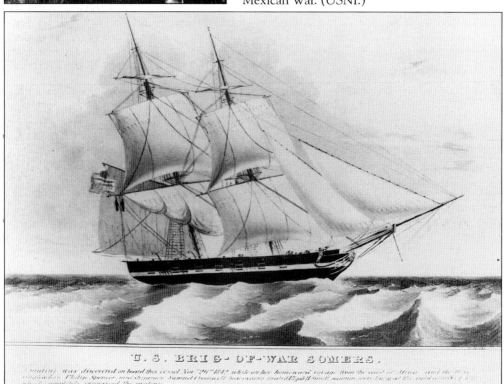

DRY DOCK NO. 1. Dry Dock No. 1 was built between 1841 and 1851, with 36-foot-deep walls built on piles driven 40 feet into the earth. This 1846 view, which may be the earliest photograph ever taken of a New York City landscape, was highlighted for publication in the *Brooklyn Eagle* newspaper. The original, unretouched photograph is in the Library of Congress. (BPL.)

THE USS ALBANY. The 20-gun, 1,064-ton sloop of war USS *Albany* was begun in 1843 and launched on June 27, 1846. It was pushed to completion in an accelerated building schedule dictated by the Mexican War (one of the few ships whose construction was sped up for the war). Toward the end of the war, it was dispatched on an intelligence mission to investigate the state of affairs in the Yucatan. (USNI.)

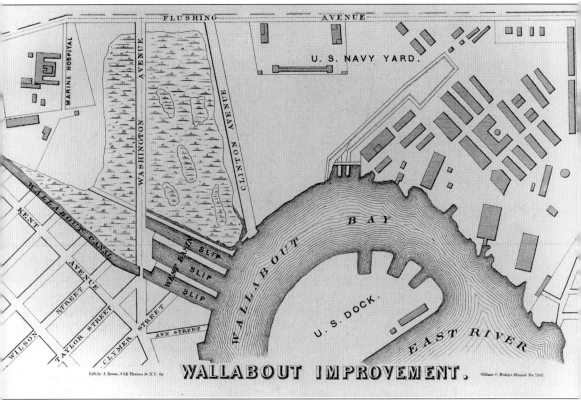

MAP OF BNY, MID-1800s. This map of BNY dates from the mid-19th century. Dry Dock No. 1 can be seen just above the "B" of the word "Bay." U.S. Dock, more commonly called Cob Dock, was an artificial island created in the middle of Wallabout Bay by merchant ships disposing of ballast stones on the site at the end of voyages. (NYHS.)

THE INTERIOR ESPLANADE OF BNY, 1851. The receiving ship *North Carolina* can be seen in the background. The pyramids are stacks of cannonballs, which were a sort of decorative object until the arsenal was needed in wartime. Old prints show large numbers of promenaders in civilian dress, which indicates that in a time when neither the City of New York or the then separate City of Brooklyn had developed their great parks, the open spaces of the navy yard were a great attraction. (NYHS.)

BNY IN THE 1850S. The three buildings in the upper left corner are officers' quarters along Flushing Avenue. These same buildings can be seen in the upper left corner of the map on page 22. (MOCNY.)

THE USS SABINE. The third ship to be started at the yard was the 44-gun frigate USS *Sabine*. Although the keel was laid in 1822, the *Sabine* was not launched until February 3, 1855. It is shown here rescuing a battalion of U.S. Marines from the sinking transport USS *Governor*, off Georgetown, South Carolina. The rescue took place with minimal loss of life while under fire in an attack on Forts Walker and Beauregard on November 2, 1861. (USNI.)

THE USS NIAGARA. The keel of the 40-gun steamer frigate USS *Niagara* was laid on October 2, 1854, and the ship was launched on February 23, 1856. Two years later, it laid the first transatlantic cable, one of the most important events in the history of modern communications. Here the Niagara is shown sailing into Trinity Bay, Newfoundland, on August 4, 1858. Note the cable-laying equipment on the stern. (MOCNY.)

MACHINE SHOP AND ENGINE ROOM, U. S. NAVY YARD, BROOKLYN.

MACHINE SHOP AND ENGINE ROOM, 1850S. The yard was one of the earliest and most important industrial sites in Brooklyn. BNY's modern technology in fields like steam engines served as a catalyst for industrial development throughout the region. (BPL.)

INTERIOR OF THE ENGINE ROOM. Despite the navy's recognition of the importance of steam power, engineering officers were treated as second-class citizens until the turn of the century, with slower promotions, fewer perks, and a lower career ceiling. (BPL.)

THE USS IROQUOIS. The six-gun steam sloop of war USS *Iroquois* was begun in 1858 and launched on April 12, 1859. It was the last ship launched at BNY prior to the Civil War. It participated in the Battle of New Orleans during the war. (USNI.)

BNY AT THE START OF THE CIVIL WAR. With the loss or neutralization of navy yards in the Southern States, BNY became more important than ever before. (NYHS.)

CONSTRUCTION AT THE START OF THE CIVIL WAR. In its first 60 years, BNY had evolved from a few buildings, with shipways consisting of planks of wood laid on the open ground, to a major industrial site with shipways built on piles, ship-houses, dry docks, and power plants. BNY was ready for war. (MOCNY.)

Two

THE AGE OF TRANSITION: 1861–1904

In the years between the start of the Civil War and the turn of the 20th century, the navies of the world experienced the most wrenching technological changes they would ever have to endure. At the start of this period, warships were primarily wind driven, wooden craft similar to vessels that had plied the oceans for thousands of years. Steam power was more than a curiosity but less than a necessity. Technological developments, including screw propellers, armor plate, and gun turrets, arrived in rapid succession, and by the dawn of the 20th century, the old days were over. The Civil War and Gen. Winfield Scott's policy of a naval blockade, which eventually defeated the South, caused a demand for a large navy, bringing about a nearly 400 percent increase in the labor force at BNY. With peace, employment was scaled back, although new buildings continued to change the face of the yard. The navy was largely neglected, however, until Presidents Benjamin Harrison and Grover Cleveland recognized the importance of a modern navy, which was confirmed by the Spanish-American War. During the transitional era, BNY built 28 major ships: one second-class battleship, one steam frigate, one protected cruiser, ten steam sloops of war, two screw steam sloops, six side-wheel double-enders, four double-turreted monitors, one torpedo boat, one screw steamer, and one steel yard tug.

BNY, TWO MONTHS AFTER UNION AND CONFEDERATE FORCES TRADED SHOTS AT FORT SUMTER. The BNY-built *Savannah* is the tall-masted ship in the center background. The average number of BNY employees and the annual payroll rose from 1,650 and $679,000 in 1861 to 5,390 and $3,735,000 in 1864 and 5,000 and $3,952,000 in 1865. At one point in 1864, 6,000 workers were employed at BNY. (BPL.)

THE USS TICONDEROGA. The nine-gun steam sloop of war USS Ticonderoga was begun in 1861 and launched on October 16, 1862. In addition to new construction, 416 commercial vessels were purchased and converted to warships. In a remarkable feat of productivity, the passenger steamer Monticello was converted in 24 hours. (USNI.)

THE USS LACKAWANNA. The nine-gun steam sloop of war USS Lackawanna was begun in 1862 and launched on August 9 of the same year. In September of the following year, BNY hosted the visiting Russian fleet, whose friendly visit helped to discourage potential British intervention on behalf of the Confederacy. The Lackawanna, meanwhile, was engaged in the Battle of Mobile Bay. (USNI.)

THE USS TULLAHOMA. The eight-gun side-wheeler USS *Tullahoma* was begun in 1863 and launched on November 28, 1863. As the navy tried to integrate new technology, it experimented with various propulsion systems. Side-wheelers became a technological dead end when it became apparent that the paddle wheel was extremely vulnerable to enemy gunfire. (MOCNY.)

THE USS MONITOR. With its all-metal construction, gun turret, screw propeller, and dozens of other innovations, the USS *Monitor,* shown here in the James River, Virginia, on July 9, 1862, was the world's first modern warship. Although not built at the Brooklyn Navy Yard, it was completed, outfitted, armed, crewed, and commissioned at the yard, making it a sort of stepchild of BNY. (LC.)

THE CONTINENTAL IRONWORKS. The Continental Ironworks, located about a mile north-northeast of BNY along Newtown Creek in Greenpoint, Brooklyn, produced seven ironclad turreted warships including the *Monitor*. Like the *Monitor*, all of these ships sailed to BNY to enter the navy. An ironclad cost the federal government about $380,000 at 1863 prices. (NYHS.)

THE USS WAMPANOAG. The screw steam sloop USS *Wampanoag* was begun in 1863 and launched on December 15, 1864. Another screw steam sloop, the *Java*, begun in 1864, became the first victim of a contract cancellation after the end of the Civil War. (USNI.)

THE USS QUINNEBAUG. The ten-gun steam sloop of war USS *Quinnebaug*, begun in 1864 and launched on March 31, 1866, is shown here during a visit to Venice. Begun shortly before the *Java*, it was the last wartime construction project to be completed. The *Quinnebaug* normally had a smokestack, not shown here, between the two lifeboats amidships (see page 2). (MOCNY.)

BNY IN MID-1866. This photograph taken at BNY in the summer or fall of 1866 shows, from left to right, the following: the *Wampanoag*; a screw gunboat of the Kansas or Cayuga class; the USS *Madawaska*, preparing for sea trials; the USS *Susquehanna*; the USS *Idaho*, with two smokestacks, which has just flunked its trials; two double-ender side-wheel gunboats; and the USS *Vermont*, which has just replaced the *North Carolina* as BNY's receiving ship. (NHC.)

THE TORPEDO BOAT ALARM. The torpedo boat *Alarm*, begun in 1872 and launched on November 13, 1873, is shown here in Dry Dock No. 1 in 1883. The *Alarm* appears to sport a ram, which became a worldwide naval fixture for 20 years after an Austro-Hungarian ship sank an Italian warship by ramming it during the Battle of Lissa in 1866. The ram proved to be a one-shot phenomenon and another technological dead end. (USNI.)

THE USS FLORIDA. The *Wampanoag*, now renamed the USS *Florida*, is shown at BNY in the winter of 1874. Shortly after the Civil War, The *Wampanoag* set speed records that were unmatched for 20 years. (SI.)

BNY IN THE 1870s. The level of ship traffic seen in the harbor on this ordinary day in the 1870s is seen nowadays only on very special occasions. The photograph shows a steam sloop in the dry dock. Cob Dock and the pre-Brooklyn-Bridge New York Harbor are in the background. (BPL.)

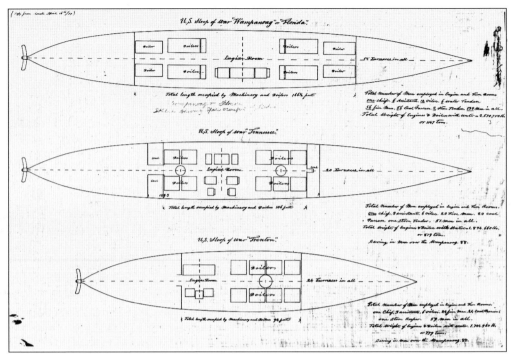

THE PROGRESS OF STEAM TECHNOLOGY. This document, labeled "Copied from Roach's April 16, 1877," illustrates the progress of steam technology. From the 1864 *Wampanoag*, top, and the 1865 *Tennessee*, middle, to the 1874 BNY-built USS *Trenton*, bottom, the power plant became progressively smaller and more efficient. Professor Roach was one of the leading mechanical engineers of the 19th century. (NA.)

BNY PRISON. During this period, many of the earliest buildings were replaced with modern ones, such as the brig, or prison, shown here. Two dry docks also were built during this period: No. 2, built between 1887 and 1890, 451 feet, 7 inches long, with a 71.5-foot beam and a 23.5-inch draft; and No. 3, built between 1893 and 1897, 612 feet long, with a 71-foot, 2-inch beam and a 26-foot, 8-inch draft. (BPL.)

THE MESS HALL. This building began as a machine shop and was converted into a mess hall. Note the manhole cover in the foreground. The Little Street Pier was built in 1890 over a portion of Brooklyn's sewer system, through which, according to legend, generations of sailors exited for a night out, just as Frank Sinatra and Gene Kelly did in *On the Town*, when liberty was denied to them. (BPL.)

BNY FROM THE EAST RIVER, 1896. The large roofed ship in this 1896 view of BNY from the East River was the yard's receiving ship, the USS *Vermont*. From 1865 to 1901, the *Vermont* provided new recruits with their first taste of life aboard a ship. (BPL.)

THE WALLABOUT MARKET. The navy sold part of BNY to the City of Brooklyn, which built the Wallabout Market with Dutch-style architecture. The Marine Hospital is in the background at the right. The navy later reclaimed and demolished the Wallabout Market in an expansion program before World War II. (BPL.)

LAUNCHING THE MAINE. The battleship USS Maine is launched at BNY. Note how much post-launch work is required: the Maine still lacks propellers, rudder, superstructure, and armament. (NHC.)

THE USS MAINE. The 6,682-ton, steel twin-screw, second-class battleship USS *Maine*, was begun on October 18, 1888, launched on November 18, 1890, and commissioned on September 17, 1895. Its captain, William Sigsbee, is in the medallion above the ship. (USNI.)

STEAM ENGINEERING BUILDING, 1890s. With the elimination of sails from most ships, engineering became a more respectable career choice. Around this time, the strategic theories of Alfred Thayer Mahan became influential, inspiring a generation of reform-minded naval officers, led by men such as Adm. William Sims. (BPL.)

THE USS CINCINNATI. The 3,183-ton, steel twin-screw, protected cruiser USS *Cincinnati* was begun in January 1890, launched on November 10, 1892, and commissioned on June 16, 1894. (USNI.)

THE USS TERROR. The double-turreted monitor USS *Terror* was launched on March 24, 1883, and commissioned on April 15, 1896. (USNI.)

THE USS MIANTONOMAH 2ND. The double-turreted monitor USS *Miantonomah 2nd* was launched on December 5, 1876, and commissioned on October 27, 1891. (USNI.)

THE USS PURITAN. The double-turreted monitor USS Puritan was launched on December 6, 1882, and commissioned on December 10, 1896. Like the Terror and the Miantonomah, the Puritan represented a transitional, pre-Mahan technology, more suited for coastal defense than for force projection, which lay at the heart of Mahan's doctrines. (NHC.)

41

THE TERROR AT SEA., This July 26, 1898 photograph shows the USS *Terror* en route from San Juan, Puerto Rico, to St. Thomas, Virgin Islands, during the Spanish-American War. The non-seagoing nature of these monitors is apparent: although the sea is not particularly choppy, the deck is awash. (NHC.)

IN DRY DOCK. The USS *Maine* is at BNY in what appears to be Dry Dock No. 2. (BPL.)

THE MAINE PASSING MORRO CASTLE AS SHE ENTERED THE HARBOR OF HAVANA

ENTERING HAVANA HARBOR. The USS *Maine* enters Havana Harbor, Cuba, on January 25, 1898. Three weeks later, the battleship blew up there, precipitating the Spanish-American War. (NHC.)

THE USS *ALGONQUIN*. The armed tug USS *Algonquin* is fitted out for war service at BNY in April 1898. The *Algonquin* served the navy from 1898 to 1947. It was successively named the *Accomac*, the *Nottoway*, and finally, more prosaically, *YTL-18*. (NA.)

BNY, APRIL 1898. The USS *New Orleans* arrived at BNY from England on its maiden voyage in April 1898. Built in England for the Brazilian navy, the *New Orleans* was purchased by the United States as war clouds gathered. To the left is the receiving ship *Vermont*. (NHF.)

JOINING THE NAVY. When war broke out, it was common for America's rich to donate their yachts for government service, often donating funds for their upkeep, as well. Here is a yacht in Dry Dock No. 1 on May 17, 1898, in the process of becoming the USS *Aileen*. (NHC.)

READY FOR ACTION. Another former yacht, the USS *Viking*, leaves BNY four days later, on May 21, 1898. The ship in the background is probably the *Aileen*, with paint job completed. (NA.)

FLEET PARADE. The war won, the fleet enjoys a victory parade in New York Harbor. The cruiser USS *Brooklyn* dominates the scene. In the background, to the right of the *Brooklyn* and just to the left of the closest tug, can be seen Grant's Tomb. (NHC.)

PEACETIME PASTIME. BNY quickly reverted to peacetime routine. Here the USS *New York* plays host to the ladies on a summer day in 1899. (NHC.)

OVERHAUL. The USS *Brooklyn* receives an overhaul in the summer of 1899. The poles wedged between the ship and the stepped sides of the dry dock keep the Brooklyn upright. (NHC.)

BNY C.1900. Small torpedo boats are tied up at BNY. The USS *Vermont*, in the background, dates this photograph to before 1902. (NYHS.)

FLOATING DERRICK. BNY's new steam-driven, steel floating derrick caught the attention of the *Scientific American* at the turn of the century. (BPL.)

In the Shoals of History:

Footnotes to the Past

In the fall of 1810, the development of a revolutionary new weapon was set back for many years by an innovative defense against it developed by Capt. James Lawrence, who became a great naval hero in the War of 1812. Steamboat inventor Robert Fulton had invented a form of torpedo launched from a harpoon gun. In March 1810, Congress appropriated funds for its development and authorized a test of the weapon at BNY. Fulton attempted to sink a sloop with his torpedo by closing within firing range in a rowboat rowed by eight men and shooting the torpedo into the hull of the ship. Lawrence, who was put in charge of defending the sloop, designed an elaborate form of anti-torpedo net which, among other things, would have decapitated Fulton had the rowboat gotten close enough to fire the torpedo. Fulton was stymied and, with his death five years later, his torpedo remained on the drawing boards.

Charles F. Guillou, a navy surgeon, served on the *Peacock* during the Wilkes Expedition and was later stationed at BNY as surgeon on its receiving ship, the *North Carolina*, in 1852. Guillou went on to become the court physician to King Kamehameha IV of Hawaii.

In 1861, BNY was attacked by a mob of more than 300 Confederate sympathizers. BNY's marines and the 14th Brooklyn Regiment were called out to quell the riot.

For a sea battle that ranks among the most significant in world history, there was a comic element to the battle between the Confederate's *Merrimac*, or *Virginia*, and the Union's *Monitor*. The gun turret of the *Monitor* failed to stop, so Lt. Samuel Dana Greene had to load and fire his guns while trapped in the spinning turret. The speaking tube connecting the turret to the captain in the pilothouse broke down, so a team of officers running back and forth kept communications open. Greene's knowledge of what was happening was severely diminished by the spinning turret and the clouds of gunpowder, but Capt. John L. Worden, seasick and temporarily blinded, was of limited help. Neither ship was damaged and Worden's temporary blindness was the only injury. The Confederates were more baffled than threatened by the *Monitor*.

In the long history of New York City con games, one of the more colorful ones involved BNY. In the 1880s, John C. Abbey, also known as "Westminster Abbey," a New York scrap dealer, claimed to have discovered at BNY the chain that George Washington strung across the Hudson River at West Point to block British ships. Although the newly discovered item bore no similarity to the real article, which was and is on public display at West Point, Abbey managed to sell links of his chain to various collectors. When he retired from the scam in the early 1900s, famed international arms dealer Francis Bannerman continued to supply the demand. Among the bamboozled were a former mayor of New York and various descendants of the ironmaster who forged the real Washington chain.

The first singing voice ever broadcast over a radio rang out at BNY on December 16, 1907, when Eugenia H. Farrar sang on the occasion of Adm. Robley D. "Fighting Bob" Evans's departure to command the Great White Fleet's around-the-world cruise. Lee DeForest, an inventor, served as the technician.

In the 19th century, the area around BNY was known as Irish Town because of its predominantly Irish residents. As BNY grew, businesses sprang up catering to sailors and workers. These ranged from Battleship Max, a uniform supplier, to numerous saloons, gambling dens, and bordellos. By the 1930s, New York City's Filipino community was located near BNY, probably because at this time, most mess hall stewards in the navy were Filipino. At a time when foreign cuisine was a rarity even in New York, restaurants offering Filipino fare in this part of town were an exception.

One of the businesses near BNY was an innocuous German-owned ice-cream parlor on nearby Graham Avenue. When World War II broke out, the FBI raided the parlor and arrested the proprietor, who had been a very effective Nazi spy by lending an ear to lonely sailors. Other German-owned waterfront shops were also closed down. The FBI had done such a thorough job of tracking down Nazi and Japanese spies that its chief, J. Edgar Hoover, was the only Roosevelt administration official to protest the President's internment of American citizens of Japanese descent in concentration camps: Hoover was confident that he had already captured every Axis spy in America.

Besides the ships built at the yard, many more ships, built at other usually private shipyards along the East Coast, sailed to BNY for inspection, outfitting, and formal acceptance by the navy at a commissioning ceremony. One such ship, the cruiser USS *Juneau*, was built at the Federal Shipbuilding and Dry Dock Company, Kearny, New Jersey, and commissioned at BNY on February 14, 1942. Five crew members got more press coverage than the *Juneau* did. They were the Sullivan brothers, handsome young Irish-Americans named Joe, Francis, Madison, George, and Allen. When Japanese warships sank the *Juneau* in November 1942, all five brothers were killed. Their saga became a successful movie, and a destroyer was named in their honor.

Among the workers at the yard was a height-disadvantaged man named Skippy LaBolla. Disadvantaged outside the Sand Street Gate perhaps, but greatly advantaged at BNY. His diminutive stature allowed him to perform repairs in parts of ships that a person of normal size would never be able to enter.

The USS *LaSalle*, launched at the yard on August 3, 1963, served as the flagship of United Nations forces during the Persian Gulf War in 1991. The *LaSalle*, like others of its kind, has a hinged mast, originally designed to allow the ship to sail under the Brooklyn Bridge.

During World War II, the Cavendish Cafeteria was a popular restaurant across Flushing Avenue from the yard. After the war, as BNY shrank, cafeteria owner Ben Eisenstadt diversified. Today, the Cumberland Packing Corporation, run by Eisenstadt's sons, is a multinational corporation that produces a number of food products including Sweet n' Low, Sugar in the Raw, and Butter Buds. With outstanding community loyalty, this multimillion-dollar corporation is still headquartered on the site of the Cavendish Cafeteria.

REINFORCING THE DRY DOCK. The modern fleet and America's new world role required the improvement and reinforcement of existing facilities, especially the dry docks, as shown here. The shipways were also remodeled, with granite girders added, to which timber ground-ways were attached. (BPL.)

MARINES AT BAT. Members of BNY's U.S. Marine contingent play baseball in a recreational field and parade ground at the yard c. 1900. A portion of the marine barracks can be seen on the left. Marine units were stationed at the yard and occasionally joined expeditionary forces when the need arose. (BPL.)

TROPHIES ON DISPLAY AT QUARTERS A. Among the trophies on display in front of Quarters A in 1904 were a breech-loading cannon from the prison hulk *Jersey*; eight breech-loading cannon from the HMS *Macedonian*, captured by the USS *United States* in the War of 1812; experimental guns from John Ericsson's ship the *Oregon*; and six guns captured from the Spanish fleet. In World War II, many of these trophies were melted down for the steel that was used to construct new warships. (MOCNY.)

RESTOCKING AT BNY. A warship receives provisions at BNY shortly after the Spanish-American War. The splash of white in the lower right of the picture is a canvas awning stretched over a portion of the deck. (BPL.)

COB DOCK FERRY, EARLY 1900S. The Cob Dock Ferry was a shuttle service that operated in the early 1900s for the crews whose ships were tied up at Cob Dock. Located on the left, Cob Dock disappeared in an expansion of the Navy Yard Basin not long after this photograph was taken. (MOCNY.)

GATE TO THE YARD. This early-1900s photograph shows the Sand Street entrance to BNY. Note the rather informal security, or lack thereof, at the entrance. Nearby, on the other side of the gate, stood a marble shaft erected to commemorate the deaths of 12 sailors in a battle in Canton, China, in 1856. (MOCNY.)

OFFICERS QUARTERS C. 1904. This photograph shows the officers quarters located along Flushing Avenue at BNY. These buildings date from before the Civil War. (MOCNY.)

MARINE BARRACKS, 1904. Judging from their lack of precision, the marines drilling in the left foreground are probably new recruits. (MOCNY.)

DRILLING IN THE PARK, 1904. These sailors drilling in BNY's park in 1904 seem no more enthusiastic about close-order drill than did the Marines in the previous picture. (MOCNY.)

STREET DRILL, 1904. Sailors drill outside of the park on Perry Avenue, one of BNY's main streets, in 1904. Perry Avenue was named after Capt. Matthew Calbraith Perry, the yard's most famous commandant. Machine Shop No. 14 is in the background. (MOCNY.)

BNY c. 1904. The road to the left leads to Quarters A. The road to the right leads past the former Naval Lyceum, which in 1904 was the commandant's office, to the East River. Pictures of BNY at the beginning of the 20th century have remarkable contrasts for such a small parcel of land, from parklike settings to grimy industrial landscapes. Industrial plants soon replaced the greensward shown here. (MOCNY.)

Three

THE AGE OF THE BATTLESHIP: 1904–1941

Historians disagree as to when the battleship era began. Some place the date as early as 1862, when the USS Monitor dueled with the CSS Virginia. Some see the Spanish-American War as ushering in the new age. Others argue for the 1904 victory of the Japanese over the Russians at the Battle of Tsushima Strait. In 1904, BNY and Philadelphia became the navy's two "battleship yards." They began launching the second generation of modern battleships, whose size, fire power, and technology dwarfed that of the Maine and others like it. Although battleship tactics dominated strategic theory throughout this period, the construction of battleships ended abruptly when Pres. Warren G. Harding initiated the Washington Naval Conference, which produced arms-control treaties that restricted the numbers of capital ships a fleet could have. This resulted in the scrapping of two battleships under construction at BNY. The 1920s and early 1930s saw little new construction, but with wars and rumors of wars in the late 1930s, the country began to stir. In the fall of 1937, a battleship keel was laid at BNY for the first time in 20 years. In 1940, both political parties promised to stay out of the war, although the candidates became ardent interventionists when the election was over. During this period, seven battleships, two heavy cruisers, three light cruisers, two destroyers, one gunboat, two coast guard cutters, a fleet collier, one covered freight lighter, and numerous smaller vessels were launched.

THE USS COLORADO IN DRY DOCK. In 1904, the USS *Colorado* was laid up for repairs in one of BNY's dry docks. (MOCNY.)

BNY RAILROAD. BNY had its own railroad system for hauling heavy materials around the yard. Here, one of the yard's trains is returning empty gunpowder canisters to a storehouse in 1904. A similar engine, built for BNY in 1918, is on exhibit today at the Rochester & Genessee Valley Railroad Museum in Rush, New York. (MOCNY.)

BNY'S FIRST STREET. This 1904 photograph shows BNY's First Street. The yard's paymaster's office is on the left. One of the shipways can be seen in the background, at the end of the street. (MOCNY.)

BNY'S THIRD STREET. This view of Third Street in 1904 shows sailors drilling, on the right. The tracks of the yard's railroad can be seen in the street. (MOCNY.)

BNY'S MORRIS AVENUE. This view of Morris Avenue was taken in 1904, looking into the yard from the East Gate. Note the bicyclist on the left. (MOCNY.)

BNY's Dock Street. This 1904 photograph shows a traveling crane on Dock Street. To the left of the crane is what appears to be two gun turrets that have been removed from a ship. The battleship USS *Texas* is in the background. (MOCNY.)

The USS Texas, 1904. This is another view of the *Texas*, taken on the same day as the previous picture. (MOCNY.)

PANORAMA. This panoramic view shows BNY in 1905 from east, upper left, to west, lower right. The Manhattan Bridge can be seen at bottom right. (MOCNY.)

THE USS CONNECTICUT. The first of the second-generation battleships, the 16,000-ton USS *Connecticut* is readied for launching on September 28, 1904. The keel was laid on March 10, 1903. (MOCNY.)

LAUNCHING THE CONNECTICUT. The USS *Connecticut* heads down the shipways during its launch on September 29, 1904. State-of-the-art in 1904, the *Connecticut* was rendered obsolete by the next generation of battleships less than ten years later. (MOCNY.)

HITTING THE WATER. As the *Connecticut* hits the water, some of the more enthusiastic members of the crowd wave their hats. One peculiarity of this photograph is that, although the ship being launched is identified as the *Connecticut*, the bunting has 48 stars, which would not have been appropriate until 1912. Either the photograph has been retouched or it is misidentified. (BPL.)

POST-LAUNCH VIEW. After its launching at BNY, the *Connecticut* is tied up at the dock before outfitting took place. It was commissioned on September 29, 1906. (MOCNY.)

OUTFITTING THE CONNECTICUT. The *Connecticut* is shown on May 1, 1906, with outfitting nearly complete. A year after this picture was taken, the *Connecticut* served as the flagship of the round-the-world cruise of Pres. Theodore Roosevelt's Great White Fleet, which asserted America's role as a world power. (USNI.)

THE USS VESTAL. The 12,000-ton fleet collier USS *Vestal*, was begun on March 25, 1907, launched on May 19, 1908, and commissioned on October 4, 1909. The *Vestal* was on Battleship Row in Pearl Harbor on December 7, 1941. A torpedo ran beneath its keel to hit the battleship *Arizona*, moored next to it. (USNI.)

DRY DOCK NO. 4. BNY's Dry Dock No. 4 was completed in 1913. It was 717 feet long with a 107-foot beam and a 32.5-foot draft. This picture was taken on November 1, 1905, shortly after construction began. The dry dock was built on a quicksand bed, which caused a collapse during construction, killing 20 workers and injuring 400. (BPL.)

THE USS NEW YORK. The 27,000-ton battleship USS *New York* was begun on September 11, 1911, launched on October 30, 1912, and commissioned on April 15, 1914. In a colorful career ranging from two world wars to an almost-war with Mexico, it was especially noted for its marksmanship at the invasion of Iwo Jima in 1945. (USNI.)

THE USS WYOMING. The 12-inch forward guns of the battleship USS *Wyoming* aim toward downtown Brooklyn on May 1, 1914. Less than two months later, Archduke Franz Ferdinand was assassinated in Sarajevo; and three months and three days after this picture was taken, the world was at war. (BPL.)

BNY, FROM THE EAST RIVER. This photograph of BNY was taken from the East River c. 1910. The brig is in the left foreground. The fleet is in port, behind Cob Dock, which later was removed to make more room in Wallabout Bay. (NYHS.)

THE USS ARIZONA. The 31,400-ton battleship USS *Arizona* was begun on March 16, 1914, launched on June 19, 1915, and commissioned on November 17, 1916. This photograph shows the *Arizona* sailing under the Brooklyn Bridge on its maiden voyage. The newly built Woolworth Building is in the background. Less than two years after the *Arizona* was commissioned, the future actor Humphrey Bogart reported for duty at BNY as a seaman. (BPL.)

THE USS NEW MEXICO. The 32,000-ton battleship USS *New Mexico* was begun on October 14, 1915, launched on April 23, 1917, and commissioned on May 20, 1918. It is shown here sailing through the Panama Canal. It served throughout World War II and was Adm. Raymond Spruance's flagship during the Battle of Okinawa when, on May 12, 1945, it was struck by a kamikaze. (USNI.)

THE USS TENNESSEE. The 32,000-ton battleship USS *Tennessee*, was begun on May 14, 1917, launched on April 30, 1919, and commission on June 3, 1920. It was the last battleship built at the yard for 20 years. Two additional battleships, the *South Dakota* and the *Indiana*, were scrapped as a result of Pres. Warren G. Harding's peace initiative. The *Tennessee* was damaged at Pearl Harbor but survived to get its revenge at the Battle of Leyte Gulf. (USNI.)

BNY'S PIER D. BNY expanded greatly during the early years of the 20th century. Eight new piers joined the Little Street Pier, including Pier D, under construction in this picture taken on May 31, 1911. (BPL.)

FLOATING CRANE. In dry dock on February 11, 1911, is this floating crane with its cross beam raised. A second shipway was constructed in 1917 and 1918 and an overhead steel runway was installed in 1921, replacing some of the duties of floating cranes such as this one. (BPL.)

HERCULES IN DRY DOCK. The floating crane Hercules is in dry dock on February 8, 1911. This view is toward the northwest. About this time, BNY survived the first attempt to close it down, when prominent Brooklyn citizens pressured the government to abandon plans to remove the yard to Communipaw, New Jersey. (BPL.)

BORDER CRISIS. On March 8, 1911, U.S. Marines march toward a tugboat that will take them from BNY to a transport bound for the U.S.-Mexican border, where one of the periodic border crises of that time had erupted. (BPL.)

DRY DOCK NO. 4, COMPLETED MAY 10, 1913. Although America's participation in World War I was not lengthy enough for major shipbuilding projects to get under way, the yard kept busy with repairs, the construction of small vessels, and the conversion to troop transports of a number of interned German commercial ships, including the *Vaterland*, renamed the *Leviathan*, one of the world's largest ships. At its peak during World War I, BNY's labor force was 18,000 people. (BPL.)

SUBMARINE CHASER. During World War I, BNY built 49 submarine chasers—small, speedy antisubmarine boats, similar to the one shown here on the Hudson River, the USS *Privateer*, which was built elsewhere in 1917. At the same time, the yard built many barges, lighters, scows, floating workshops, and floating derricks. The larger ship shown here is probably the USS *San Diego*, sunk by a mine off Long Island in 1918 and now a popular diving site. (NHC.)

BNY BOAT SHOP, EARLY 1930S. This view shows the three-story boat shop at BNY in the early 1930s. In addition to the manufacture of assorted goods, BNY housed the navy's chemical, electrical, and radio laboratories and a merchant marine training center during this period. The Navy Electrical School was located at BNY until 1916 when it was divided into several more specialized schools. BNY also housed the Sound Motion Picture Technicians School and the Gyrocompass School during this period. (BPL.)

GARBAGE HAULER. A truck hauls the yard's garbage to a New Jersey pig farm. By this time, BNY had become so large that gun salutes honoring approaching ships could no longer be heard in the commandant's office; so they were reported by telephone and, later, broadcast by loudspeaker. No ships at all were launched at BNY between April 30, 1919 (the *Tennessee*) and April 25, 1929 (the *Pensacola*). (BPL.)

THE USS PENSACOLA. The 10,000-ton heavy cruiser USS *Pensacola* was begun on October 27, 1926, launched on April 25, 1929, and commissioned on February 6, 1930. With battleship construction halted by Pres. Warren G. Harding's disarmament triumph, countries concentrated on smaller ships such as cruisers. Shown here delivering troops to Midway Island shortly after the eponymous battle, the *Pensacola* was damaged in the Battle of Tassafaronga on November 28, 1942. Following the war, the ship was sunk after surviving an atomic bomb test off Bikini Atoll on July 1, 1946. (NA.)

FLOTILLA OF SUBMARINES. This c. 1930 photograph show a flotilla of submarines at BNY for provisioning. Although one of the trophies at BNY was the navy's first submarine, the "Intelligent Whale," built in 1864 in New Jersey, the yard never developed an expertise in submarine construction. (NHC.)

THE USS NEW ORLEANS. The 10,000-ton heavy cruiser USS *New Orleans*, was begun on March 14, 1931, launched on April 12, 1933, and commissioned on February 15, 1934. It was one of eight BNY-built ships attacked at Pearl Harbor. The others were the *Arizona, Tennessee, Helena, Honolulu, Hull, Dale* and *Vestal*. The *New Orleans* was later damaged at the Battle of Tassafaronga on November 28, 1942.(USNI.)

HEAVY CRUISER AND DESTROYERS. At BNY on December 9, 1931, were the heavy cruiser USS *Northampton*, right, sunk at Tassafaronga in 1942; and the destroyers USS *Barry*, foreground; USS *Gilman*, left; and USS *Wickes*, right. By 1931, Japan and China were skirmishing, the world economy was slipping into a depression (U.S. unemployment had risen from 8.7 percent to 15.9 percent during the year), and the Nazi party was beginning to get significant support in Germany. (BPL.)

THE USS HULL. There were few new ships built during the Depression and those tended to be smaller vessels. The 1,395-ton destroyer USS *Hull* was begun on March 7, 1933, launched on January 31, 1934, and commissioned on January 11, 1935. The *Hull* survived Pearl Harbor but was lost in a typhoon in the western Pacific in December 1944.(USNI.)

THE USS ERIE. The 2,000-ton gunboat USS *Erie* was begun on March 17, 1934, launched on January 29, 1936, and commissioned on July 1, 1936. Although new construction at BNY was modest, the yard housed the Central Drafting Office from 1929 to 1937, which had responsibility for designing all of the navy's surface fleet except for aircraft carriers. Four battleships, 11 cruisers and more than 50 destroyers were designed at BNY during this period. (USNI.)

THE ALEXANDER HAMILTON. The coast guard cutter *Alexander Hamilton* was begun on September 11, 1935, launched on January 6, 1937 and commissioned on March 4, 1937, the day of Pres. Franklin D. Roosevelt's second inauguration. Adolf Hitler had been in power for a month longer than Roosevelt and had issued his anti-Semitic Nuremberg Laws about the time that the *Hamilton* was begun. Japan was building its fleet. Italy was expanding its empire. The Soviet Union was purging its leadership. The world had become a scary place. (USNI.)

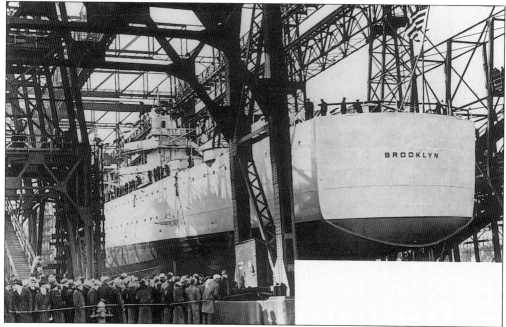

THE USS BROOKLYN. The 10,000-ton light cruiser USS *Brooklyn* entered the fleet on September 30, 1937. Its keel had been laid on March 12, 1935, and it is shown here on the day of its launching, November 30, 1936. (BPL.)

LAUNCH OF THE BROOKLYN. The USS *Brooklyn* greets the water for the first time. By this time, Adolf Hitler had marched his troops into the Rhineland, in violation of the Versailles Treaty. (BPL.)

ON THE EAST RIVER. This photograph shows the USS *Brooklyn* on the East River. The *Brooklyn* took part in several of World War II's most famous battles, including the Normandy invasion. It was damaged by a mine during the invasion of Sicily. (BPL.)

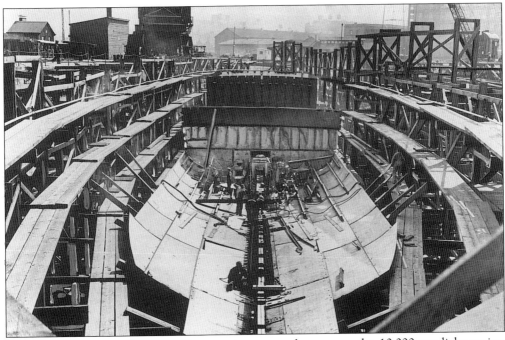

CONSTRUCTING THE CRUISER. Construction got under way on the 10,000-ton light cruiser USS *Honolulu* shortly after the keel was laid on September 10, 1935. (BPL.)

THE USS HONOLULU. The USS Honolulu was launched on August 26, 1937, and commissioned on June 15, 1938. By the time the *Honolulu* joined the fleet, Adolf Hitler had seized Austria and was demanding the Sudetenland from Czechoslovakia and Japan had invaded China. The *Honolulu* fought in the Pacific in battles ranging from the Aleutian Islands in Alaska to Guadalcanal in the South Pacific. It was damaged in the Battle of Kolombangara, July 13, 1943. (USNI.)

THE USS HELENA. The 10,000-ton light cruiser USS *Helena* was begun on December 9, 1936, launched on August 27, 1938, and commissioned on September 18, 1939, just 18 days after Hitler had invaded Poland and 15 days after England and France declared war. The *Helena* was damaged at Pearl Harbor and fought in many of the battles around Guadalcanal before being sunk in the Battle of Kula Gulf, July 6, 1943. (USNI.)

FLEET MANEUVERS. As war approached, maneuvers were no longer just war games. Here, the USS *Cincinnati* leaves BNY to join 48 other warships participating in fleet maneuvers off Norfolk, Virginia, on January 3, 1939. (BPL.)

NEW MAST. Repair and modernization were major activities of BNY during the 1930s. Here, the *Arizona* sports a new mast structure, replacing the "cage" masts seen in earlier photographs. (USNI.)

JOB OPENINGS. With the war came the end of the Great Depression. Here men are lining up for job openings at BNY on May 24, 1940, as the yard announced an accelerated construction schedule and made plans for a second shift. On the same day that this picture was taken, the defeated British Expeditionary Force was lining up on the beaches of Dunkirk, France, to escape Hitler's blitzkrieg. (BPL.)

TRADING DESTROYERS FOR BASES. In a move that edged the United States closer to war, Pres. Franklin D. Roosevelt traded 50 old four-stack destroyers to Great Britain in exchange for military bases on British-owned islands in the Caribbean. Here, eight of the old destroyers are docked at BNY shortly before the transfer. Two modern destroyers, not part of the deal, are docked in the background. (BPL.)

KEEL-LAYING CEREMONY. Rear Adm. Clark Woodward, commandant of BNY, drives the first rivet into the USS *Missouri* during keel-laying ceremonies on January 6, 1941. Even with accelerated, round-the-clock building schedules, it took three years to build a battleship. (NHC.)

BATTLE STARS. The 35,000-ton battleship USS *North Carolina* was begun on October 27, 1937. The *North Carolina* won seven battle stars in actions ranging from Guadalcanal to Okinawa. (USNI.)

81

THE USS NORTH CAROLINA. The *North Carolina* was launched on June 13, 1940, shortly before France surrendered to Germany. Another ten months of post-launch construction and outfitting was required before the *North Carolina* was commissioned on April 9, 1941. By that time, Adolf Hitler's invasion of the Balkans was under way, with his attack on the Soviet Union to follow shortly. (NYHS.)

FLAG RAISING. The flag was raised on the *North Carolina*, immediately after its launch. This battleship was the victim of the most successful submarine attack in history on September 15, 1942, when a lone Japanese submarine fired a single spread of six torpedoes, sinking an aircraft carrier and a destroyer and severely damaging the *North Carolina*. (BPL.)

Four

THE AGE OF TOTAL WAR: 1941–1945

In 1941, preparation for war began in earnest at BNY and the Battleship Era ended forever on December 7, 1941, when Japanese bombers and torpedo planes blasted America into World War II . Major construction projects were accelerated. With America's entry into the war, BNY boomed. In the allocation of resources, the United States gave priority to building the navy over building the army, which many strategists believe resulted in an army that was too small and a navy that was too large for the country's needs. Because of the demand for new ships and repair facilities, BNY's new facilities were put to use before the paint was dry. Although two battleships, two aircraft carriers, eight tank-landing ships (LSTs), and floating workshops were launched and commissioned at BNY during the war, BNY's primary responsibility was ship repair. More than 5,000 bomb- and torpedo-damaged ships from many Allied nations steamed into New York Harbor for repairs, and an additional 250 ships were converted from civilian to wartime use. By the end of the war, BNY was the largest shipyard in the world, employing over 75,000 workers, with a monthly payroll of between $15 million and $16 million. To put the number of workers into perspective, at that time 75,000 people constituted more than one tenth of one percent of the total labor force of the entire country, and they all worked on a parcel of land that could be fit four times into New York City's Central Park.

ATTACK ON PEARL HARBOR. This photograph of the Pearl Harbor attack was taken from a Japanese bomber. The BNY-built ships *Vestal* and *Arizona* are tied up side by side in the foreground, the second and third ships from the left. Directly across the harbor and a little farther to the right than the bow of the *Arizona*, the BNY-built cruiser *Helena* is burning from a bomb hit. Eight of the more than 90 ships at Pearl Harbor were built at BNY and many more were designed there. (NHC.)

NEW SYMBOL. This photograph shows ships at BNY viewed from under the new symbol of the yard, the huge hammerhead crane. This crane was capable of lifting 350 tons (some sources say 450 tons) over a radius of 115 feet. The yard's motto, "Service to the Fleet," was emblazoned on the side of the crane. (USNI.)

LANDING SHIP, TANK (LST) 311. Shown here somewhere in England prior to the Normandy Invasion is LST *311*. This ship was one of eight of its kind built at BNY on an accelerated basis. The keels for LSTs *311*, *312*, *313*, and *314* were laid on September 7, 1942, and the ships were launched on December 30, 1942. (USNI.)

LAUNCHING OF LST 312. The launching of LST 312 took place on December 30, 1942. Some of the LSTs were built in the newly completed Dry Dock No. 5. (NYHS.)

LSTs 317 AND 318. LSTs 317 and 318, together with LSTs 315 and 316, comprised the second group of LSTs built at the yard. They were begun on October 14, 1942, and launched on January 23, 1943. Of the eight LSTs built at BNY during World War II, four were sunk in action in the European or the Mediterranean theaters of operation. (USNI.)

BLUEPRINT ROOM. This photograph shows one corner of the blueprint room in the drafting department of BNY. Tons of blueprints were required for each ship built. It took 430,000 man-days to design a battleship. The amount of work one man can accomplish in one eight-hour day is one man-day. Theoretically, this means that one man could design a battleship in about 1,178 years if he did not take any vacations. Alternatively and equally theoretically, if the entire staff of BNY could have been put to work on the project, a battleship could have been designed in less than six days. (NYHS.)

PATTERN SHOP. At BNY's pattern shop, a wooden pattern for a steel gear is being sculpted. The pattern will be used to make the mold into which molten steel is poured to make the gear that runs a warship. (NYHS.)

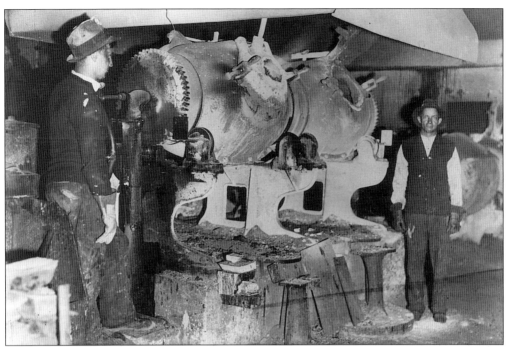

FOUNDRY. This photograph shows BNY employees working in the foundry. Spread across BNY's 219 acres and more than 300 buildings was a vast industrial plant making dozens of products. The yard was an industrial conglomerate before that term was coined. (BPL.)

BLAST FURNACE. BNY's blast furnace worked 24 hours a day. Here a piece of structural steel is fed into the furnace to be melted down. As part of the war effort, Boy Scouts collected scrap metal from Brooklyn neighborhoods and delivered it to the BNY scrap yard. That metal came to the blast furnace. (NYHS.)

Dateline: January 31, 1941

Had commandant Rear Adm. Clark Howell Woodward taken stock of the state of the yard on January 31, 1941, he would have had every reason to feel confident in BNY's ability to meet the challenges that were coming. If Woodward looked out of the windows of Quarters A, he would have seen the trophies of America's earlier triumphant wars located in front of his home. The country had survived earlier dangers and it would survive the current ones.

The Great Depression was dying hard. Unemployment in the United States in early 1939 had been as high as it was in early 1932. With Europe rearming and then at war, unemployment began to shrink, as orders for military goods flooded America's factories. However, in early 1941, nearly 15 percent of the labor force was still jobless nationwide. BNY's activity had picked up only recently. Because its production was not for export, the yard's recovery depended on the buildup of American military might, and that had only just begun.

As Woodward was driven to his office at the new administration building in one of the 5 staff cars assigned to BNY, he would have viewed the construction program that was under way. Two new dry docks, No. 5 and No. 6, both of them 1,067 feet long, were nearing completion. A huge hammerhead crane was being installed. An annex to the yard was being built in Bayonne, New Jersey, where taller ships would have their masts removed before sailing under the bridges to Brooklyn. Dry Dock No. 7 would be built in Bayonne. Additional annexes were being created in Maspeth and Flushing, in the borough of Queens, where landing craft engines would be repaired and reconditioned. A supply annex was established in Long Island City, also in the borough of Queens.

The number of workers at the yard had jumped from an average of 13,711 in 1940 to 17,182, with a monthly payroll of $1.7 million on January 31, 1941. Still, in some ways it remained a quiet Victorian-era post: the security force consisted of only 65 men—a senior watchman, 6 principal watchmen, 32 watchmen for the gates, 22 ship keepers, and 4 watchmen for the supply room; and the fire department was a total of 5 marines, with an additional 16 on call in emergencies.

As the commandant reached his office, he would have checked the progress reports on ship construction. For the first time in its history, BNY had three battleships under construction simultaneously. Woodward would have also reviewed the reports of the other manufacturing divisions: engine parts, torpedo compressors, turbine blades, exhaust headers, fire control instruments, valves, flanges, metal blocks, flags and bunting, hammocks, awnings, tablecloths, mattress covers, pillowcases, lifeboats and life rafts. He would have checked on the Applied Science Laboratory, where 900 people were testing navigational equipment with modern metallurgical techniques.

Woodward would certainly have spent time on accounting matters, as an audit was being taken on January 31, 1941. When the numbers came in a few months later, the value of BNY's physical plant was appraised at a preinflation $38.5 million. Work in progress was valued at $54.9 million. The value of ships stores on hand for supplying the fleet was appraised at $19.6 million.

Perhaps the commandant met with his opposite number commanding the Atlantic Fleet, whose headquarters were located just across the Brooklyn Bridge over the post office at 90 Church Street in Manhattan. Had the commander of the Atlantic Fleet asked Woodward if the yard was ready, the response would have been affirmative.

16-INCH GUNS. The main armament for the battleship USS *Iowa*, 175-ton 16-inch guns, arrive at BNY for installation. Construction began on the Iowa on June 27, 1940. It took 3.3 million man-days of post-design construction to build a battleship, so that after our theoretical man spent 1,178 years designing the battleship, he could spend the next 9,041 years building it. (NYHS.)

LAUNCH PREPARATIONS. Preparations are made on August 26, 1942, for the *Iowa* launching ceremonies scheduled for the following day. (NYHS.)

THE CHRISTENING. The wife of Vice Pres. Henry A. Wallace christens the 45,000-ton USS Iowa on the day of its launching, August 27, 1942. The Iowa saw action in both the Atlantic and the Pacific Oceans, ferried Pres. Franklin D. Roosevelt to the Teheran Conference, and served as Adm. William Halsey's flagship for part of the war. (BPL.)

AFTER THE LAUNCHING. This photograph shows the USS *Iowa* immediately after its launching from the shipways on the right. The tugs are ready to move the *Iowa* to a pier. There, the ship was completed before it was commissioned on Washington's Birthday, February 22, 1943. (NYHS.)

THE USS IOWA, 1945. The Iowa was brought out of retirement in the 1980s. On April 19, 1989, an explosion in turret 2, the second turret from the bow, killed 47 crew members. The cause of the explosion is a mystery to this day. (USNI.)

STEEL KEEL. A new ship takes the place of one that has just been launched. A dockside crane lifts a huge portion of a steel keel onto the shipways, as another capital ship is begun. (NYHS.)

Landing Craft Vehicle, Personnel (LCVP). At BNY, a landing craft vehicle, personnel (LCVP) is loaded aboard a navy attack transport for an invasion. The LCVP was capable of transferring 32 men or a jeep and a trailer from an oceangoing transport directly onto an enemy-held beach. (BPL.)

Flag Loft. In BNY's flag loft, hundreds of women made American flags, signal flags, streamers, emblems and all other flags and bunting required by a well-dressed navy. (NYHS.)

WOMAN'S WORK. With men at war, women filled many jobs at the yard. They worked not just in the divisions that had always been dominated by women, such as sewing flags, but also in traditionally male jobs. During the war, BNY won six navy "E for Excellence" awards. (BPL.)

HIRING PREFERENCE. Wounded veterans were giving hiring preference. Here, Richard Bright of Harlem is preparing helmets. Immediately after Pearl Harbor, when Hawaii was expecting an invasion at any moment, hundreds of BNY workers volunteered to go to Hawaii to perform emergency repairs on the stricken fleet. (NYHS.)

LUNCH BREAK. As the noon whistle blows, work crews leave the shipways for lunch. (NYHS.)

GATEWAY AT NOON. Workers pour out of the Sand Street Gate for a fast lunch before heading back to work. (NYHS.)

INSIDE MEAL. In an effort to reduce the amount of time lost to lunch, BNY set up lunch counters within the yard where a complete lunch was served in a paper container. (NYHS.)

STUDENT WELDERS. Student welders take a lunch break. In addition to mixing genders and races, the yard also became multinational. Many Allied warships were repaired at BNY, and a portion of the receiving station was turned over to the Royal Navy. (NYHS.)

BIG YARD. By the start of the war, BNY contained more than 300 buildings, 9 piers, 2 shipways, 7 dry docks, 24 miles of railroad tracks, 5-plus miles of paved road and 6-plus miles of paved sidewalk on 219 acres of land, fronting an additional 72 acres of water. (USNI.)

RAVES FOR REPAIRS. BNY developed a reputation for fast and innovative repair work. In 1943, the USS *Holder*, right, lost its bow and the USS *Menges*, left, lost its stern to Nazi torpedoes in the Mediterranean. Both of these destroyer escorts were towed to BNY, where the stern of the *Holder* was welded to the bow of the *Menges*. Rechristened the *Menges*, since more of it remained, the ship returned to war. (NYHS.)

THEN A SENATOR. At BNY for the January 29, 1944 launching of the 45,000-ton battleship USS *Missouri* were Missouri Sen. Harry S. Truman and his daughter and wife. By the end of the year, Senator Truman was Vice President-Elect Truman. (BPL)

DOWN THE SHIPWAYS. The *Missouri* heads down the shipways. Begun on January 6, 1941, nearly a year before the United States entered the war, the *Missouri* was commissioned on June 11, 1944, when the war had little more than a year left to run. (NHC.)

THE 'MIGHTY MO.' The *Missouri* hits the water for the first time. The "Mighty Mo" reached the Pacific in time to participate in the last campaigns of the war, including Iwo Jima. It was the last battleship to enter the fleet of any nation. The battleship era, which began with the construction of the *Maine* at BNY, ended at the same spot 50 years later. (BPL.)

THE USS HARLAN DICKSON. The destroyer USS *Harlan Dickson* crosses New York Harbor in February 1944, heading to BNY from its place of construction at the Federal Shipbuilding and Dry Dock Company in Kearny, New Jersey. Like scores of ships built at civilian shipyards in the area, the *Dickson* was accepted into the navy at BNY. (NHC.)

GUN SHOP. Workers at BNY's gun shop repair the firing mechanism of a 3-inch gun before its installation on a warship or armed freighter. (NYHS.)

AERIAL VIEW. This aerial view of BNY was taken on March 9, 1944. The *Missouri* is in the center of the picture. The aircraft carrier at the bottom of the picture is probably the *USS Bennington*. The tall white building at upper left is the 14-story headquarters of the yard. (NHC.)

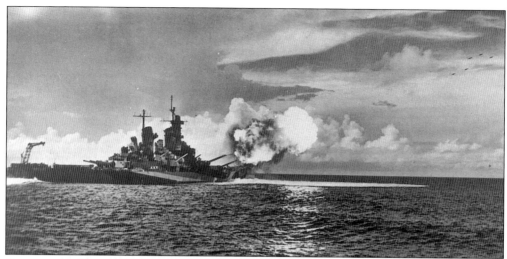

SHAKEDOWN CRUISE. The *Missouri* tests its guns off Coney Island during its shakedown cruise in mid-1944. Just visible in the upper right are the six one-ton shells fired by the *Missouri's* main battery, moving so slowly that their progress can be tracked across the sky. (NA)

THE USS BENNINGTON. The 27,100-ton USS *Bennington* was the first aircraft carrier to be built at BNY. Begun on December 13, 1942, it was launched on February 26, 1944, and commissioned on August 6, 1944. As it became clear that the day of the battleship was over, battleship contracts were canceled and replaced with aircraft carrier orders. (USNI.)

THE FAMOUS FRANKLIN. One of the most famous repair jobs at BNY was performed on the aircraft carrier USS Franklin in 1945. When the Franklin was struck by a kamikaze 50 miles off the coast of Japan on March 19, 1945, the ship was saved in an extraordinary feat of seamanship which won its crew members 2 Medals of Honor, 19 Navy Crosses, and hundreds of lesser commendations. Making only emergency repairs en route, the crew sailed the Franklin eastward halfway around the world to BNY. (USNI.)

THE JOURNEY'S END. The damaged Franklin approaches BNY on April 26, 1945. Its long journey at the very end of the war oddly mirrors a similar epic at the very beginning. In the earlier case, the light cruiser USS Marblehead was severely damaged off Borneo on February 4, 1942. With its bow submerged, the Marblehead sailed westward to BNY, arriving on May 4, 1942. (NA.)

BNY AT SUNDOWN. The lights go on. The shifts change. The work continues. (NYHS.)

USS BON HOMME RICHARD. The 27,100-ton aircraft carrier USS *Bon Homme Richard* was begun on February 1, 1943, launched on April 29, 1944, and commissioned on November 26, 1944. It was the last BNY-built ship to enter the fleet before the end of the war. (USNI.)

THEY MADE THEM TOUGH IN BROOKLYN. The aircraft in the center of the picture is a bomb-laden kamikaze that attacked the *Missouri* on April 28, 1945 off Okinawa. The Missouri shrugged off the attack. The kamikaze pilot was not able to do the same. (NHC.)

THE SHIPS ARRIVE. The *Missouri*, in the foreground, and the *Iowa* arrive in Tokyo Bay on August 29, 1945 for the surrender ceremony that took place four days later. (NHC.)

THE USS NEW MEXICO. The *New Mexico*, modernized, battle-scarred, but ready for action, also arrives off Tokyo. Mount Fuji is in the background. (NHC.)

THE SURRENDER. The sky over the *Missouri* darkens with warplanes at the moment the Japanese nation surrendered to Gen. Douglas MacArthur. The war that began with the sinking of one BNY-built battleship ended in triumph on the deck of another. (NA.)

Five

THE END OF AN AGE:
1945–1966

With the coming of peace, BNY shrank rapidly. Twelve ship orders were canceled (it took 12 years to negotiate the contract settlements). Much of the activity in the late 1940s involved the winterizing of warships destined for the mothball fleet. With the start of the Korean War, the BNY dewinterized some of the same ships. New ships were built. BNY built a series of aircraft carriers and then a group of attack transports. In all, the yard completed six aircraft carriers and six amphibious transports after the war. But the days of the yard were numbered. It never built submarines, and it never worked with nuclear ships. Modern ships were so large that many of them could not get under the Brooklyn Bridge and the Manhattan Bridge to get to the yard. The coming of the Vietnam War put an end to the yard as the Kennedy and Johnson administrations sought budget-cutting measures to finance their war. On June 25, 1966, the flag was lowered for the last time and the 165-year old institution ceased to exist, although some of the ships built at the yard continued to serve the country in 1999.

THE USS FRANKLIN D. ROOSEVELT. The 45,000-ton aircraft carrier USS *Franklin D. Roosevelt* was begun on December 1, 1943, launched on April 29, 1945, and commissioned on October 27, 1945. Launched as the war was ending, the ship was named to honor the President. On July 21, 1946, it became the first aircraft carrier in history to have a jet land on its deck. It was one of the first aircraft carriers with armored decks, and was too large for the Panama Canal. (BPL.)

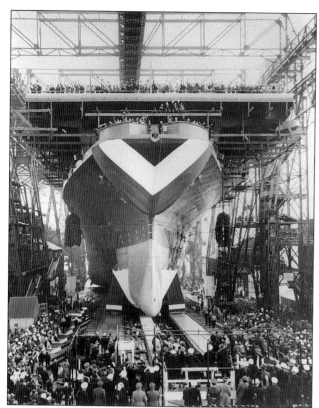

THE USS KEARSARGE. The 27,100-ton aircraft carrier USS *Kearsarge* is shown here, just prior to its launching on May 5, 1945. It launch was the second of a major warship at BNY in a week. Begun on March 1, 1944, the *Kearsarge* was the last ship launched at BNY during World War II. (NYHS.)

THE COMMISSIONING. The Kearsarge was commissioned on March 2, 1946, after the war ended. In the ten years following the Kearsarge's commissioning, only one other BNY-built ship joined the fleet. (USNI.)

THE USS ORISKANY. The 37,000-ton aircraft carrier USS *Oriskany*, begun on May 1, 1944, and launched on October 13, 1945, was not commissioned until September 25, 1950. The *Oriskany* was the last ship at BNY to be launched the old-fashioned way: stern first, sliding down the shipways into the water. Thereafter, ships were built in dry dock, which was then flooded for the launching. (USNI.)

REPRISAL SUSPENDED. What would have become the aircraft carrier USS *Reprisal* was the first construction project at BNY to be suspended with the end of hostilities in August 1945. (NYHS.)

IN TO MOTHBALLS. The USS *New Jersey* sails into BNY on February 3, 1948, to be prepared for the mothball fleet, a vast legion of unused ships parked in the backwaters of America for use in emergencies. Mothballing required, among other things, the sealing of hatches, the removing of secondary armament and radio and radar antennae, and the covering of gun tubs with igloo-shaped aluminum sheds. (BPL.)

OUT OF MOTHBALLS. The Korean War brought many of the ships back out of the mothball fleet. Here, the *New Jersey*, assisted by 16 tugboats, returns to BNY on November 21, 1950, for reconditioning in order to rejoin the active fleet. (BPL.)

THE ADMIRALS' REVOLT. The USS *Missouri* returns to New York for Navy Day, October 27, 1948. At this time, American defense strategy called for almost total reliance on the atomic bomb. In the 1949 "Admirals' Revolt," several officers sacrificed their careers to preserve naval aviation by criticizing the morality of nuclear strategy. As with Gen. Douglas MacArthur in Korea in 1951, military officers thought that the Constitution required them to testify honestly to Congress, while Pres. Harry S. Truman demanded absolute obedience to the commander in chief. Truman won. Vietnam followed. (BPL.)

THE USS WASP. The bow of the aircraft carrier USS *Wasp* was shorn off in a collision with the destroyer USS *Hobson* in 1952. In a feat reminiscent of the *Holder-Menges* repair, the bow of the *Wasp* was replaced by the bow of the USS *Hornet* at BNY. (USNI.)

POSTWAR CONSTRUCTION. This photograph shows construction under way on a postwar aircraft carrier, probably the 56,300-ton USS Independence, launched on June 6, 1958, and commissioned on January 10, 1959. (USNI.)

THE USS SARATOGA. A 65-ton high-pressure boiler is lowered into the 56,000-ton aircraft carrier USS *Saratoga*, begun on December 16, 1952, and launched on October 8, 1955. At the time of its construction, the *Saratoga* was the largest aircraft carrier in the world. (BPL.)

JET-PROPELLED CHANGES. As jets replaced propeller-driven airplanes, aircraft carrier technology required changes as well, such as the addition of catapults and angled decks. The existing fleet was converted. Here is the *Oriskany* before conversion. (USNI.)

CATAPULTS AND ANGLED DECKS. This photograph shows the *Franklin D. Roosevelt* after conversion. Note the catapults protruding from the bow and the angled deck to the right. The first angled-deck conversion was performed at BNY on the USS *Antietam* in December 1952. Tests begun on January 12, 1953, proved the value of angled decks. (USNI.)

THE USS MOUNT BAKER. The USS *Mount Baker* undergoes repairs in dry dock at BNY in the late 1950s or early 1960s. (USNI)

FABRICATING ROOM. In the fabricating room at BNY, templates were prepared from plans. (USNI.)

Dateline: September 25, 1966

One of the peculiarities of the Brooklyn Navy Yard is that, although there is very little written about the history of its operation, it is the subject of one of the best studies of the politics of military base closings ever written. As her 1974 Ph.D. thesis from the University of Illinois at Urbana-Champaign, Lynda Tepfer Carlson wrote *The Closing of the Brooklyn Navy Yard: A Case Study in Group Politics*. In her work, she examines the closing from the points of view of the Defense Department, private contractors, the navy, the labor unions, Congress, and New York City.

With the escalation of the Vietnam War, Pres. Lyndon Johnson promised "guns and butter," a war without any rationing or any cutbacks in the government's social programs. Besides causing the inflation of the 1970s, this policy required a detailed search for areas of the budget that could be cut. The task to find cuts in the defense budget was given to Robert Strange McNamara, secretary of defense. Private military contractors, who resented the government-subsidized competition, opposed the navy yards in principle. Since defense contractors were major contributors to the Democratic party, their opposition was relayed to the White House, even if the ultimate decision may have been based on other factors.

The navy had been having problems with BNY for some time. Crime had risen in the neighborhood. There were charges of corruption and low productivity. The physical plant was tired—although, ironically, the oldest dry dock in the yard was also the soundest: one writer described Dry Dock No. 2's tiered walls as resembling the ruined seats of a Roman amphitheater, and several of the other dry docks were just as bad. The size and location of BNY was not ideal. A large part of modern naval technology was outside of its expertise. For these reasons, BNY was rated a "non-core," and therefore expendable, installation, along with another 5 of the navy's 11 government-owned yards: Boston, Mare Island, Philadelphia, Portsmouth, and San Francisco. Pearl Harbor, Long Beach, Puget Sound, Charleston, and Norfolk were considered the hard-core installations, essential to the navy's mission.

Nevertheless, when McNamara proposed closing the non-core yards, the navy opposed the closings, claiming that the government-owned yards produced ships of higher quality because navy officers had closer supervision of the construction process. Because such claims could not be quantified by the sort of statistical analysis to which he was devoted, McNamara ignored these arguments; and BNY, which had already shrunk from a payroll of 12,449 on July 13, 1962 to 10,593 on March 31, 1964, was slated for the scrap heap of history.

The unions and other fraternal and labor groups held rallies to oppose the closings, but they refused to address the issues of productivity and corruption and did not bother to counter the arguments of McNamara's whiz kids with cost-benefit analyses of their own. By refusing to support their claims or promise improvements, labor groups relied solely on emotional arguments, allowing McNamara to win the analytical arguments. And reform clearly was needed. Carlson quotes one worker who explained that every morning, laborers raced to the bathroom after punching in, so that they could claim a toilet stall in which to hide out with cigarettes and magazines until lunchtime. It is horrifying to read of such an action barely 25 years after hundreds of BNY workers volunteered for war zone duty.

Once the decision for closure was made by the administration, the Brooklyn Navy Yard was doomed. Because the Democratic party dominated New York City, the Republicans had no incentive to fight the closing, intent on saving bases in Republican districts. Local Democrats were disinclined to fight their own President, especially after his stunning victory over Sen. Barry Goldwater, so they contented themselves by making promises with no intention of keeping them. Of the local newspapers, only the Republican *Journal-American* and *Herald Tribune* opposed the closing; the *New York Times*, then as now the voice of the Democratic party, endorsed it.

On August 14, 1965, the amphibious transport USS *Duluth* became the last ship to be launched at the yard. It was registered as the 84th ship to be launched at the yard since 1820—although that depends on what the meaning of the word "ship" is: there were, of course, many more than 84 vessels built at the yard over its 165-year history. The last ship overhaul was completed when the aircraft carrier USS *Intrepid* left the yard on the 20th anniversary of the end of World War II, September 2, 1965. The *Intrepid* is now a museum located on the Hudson River side of Manhattan. At 12:58 p.m. on June 25, 1966, with Rear Adm. J.H. McQuilkin, the last commandant of the yard, presiding, S1c. Stephen Bovey of South Brooklyn lowered the flag for the last time.

RUSH OVERHAUL. The guided missile destroyer USS *Rush* was overhauled at BNY in the early 1960s, seven days into the fleet rehabilitation and modernization program, FRAM. At this stage of the overhaul, BNY's work on the *Rush* showed the fastest progress of any ship in the FRAM program in any shipyard in the country. (USNI.)

BNY, LATE 1950s. This photograph shows BNY as seen from the East River in the late 1950s. An aircraft carrier is in dry dock, and several smaller vessels are in port. (USNI.)

THE USS CONSTELLATION. The USS *Constellation* was launched on October 8, 1960. On December 19, 1960, while the ship was being outfitted, a fire broke out and burned for 12 hours. Fifty workers were killed, and 150 were injured. The ship sustained $75 million worth of damage, setting back its completion by one year. Jets from the *Constellation* responded to the Gulf of Tonkin incident in 1964, initiating the Vietnam buildup. (USNI.)

LAST SHIP. The last ship built at the yard was the USS *Duluth*, a transport known as an assault personnel, dock. Mrs. Bruce Solomonson, a daughter of Vice Pres. Hubert H. Humphrey, christened the ship at its launch on August 14, 1965. Sen. Robert F. Kennedy and representatives from the New York City congressional delegation were invited to the launch but avoided it to escape the anger of betrayed workmen who had lost their jobs. (USNI.)

EPILOGUE

And so it is gone. What was once a proud symbol of the strength and power of democracy was sold to the City of New York for $22.4 million to serve as an industrial park. A modest amount of civilian ship repair is conducted there, but the city turned its back to the sea many years ago and has long neglected one of the best harbors in the world. Parts of the yard are used for light industry. Part of it is used as a storage area for city-owned vehicles. And another portion is used as an impound lot for towed and confiscated automobiles. A group of investors are currently negotiating to turn a portion of the yard into a movie studio. The yard is being used for purposes appropriate for a consumer society. Perhaps it is not wise to dwell too long on a lost era, a time when the public thought that virtue, courage, and community effort were priceless, not worthless. We are stuck where we are and must make the best of it. Flags are furled and ships depart. Iron rusts and mortar crumbles. Soon, all that is left are fading memories of a different world, one where everyone from the President of the United States to the newest immigrant knew his duty to the country and, striving for honor, achieved eternal glory.

BNY, 1999. This view from the Williamsburg Bridge shows the Brooklyn Navy Yard at the end of the 20th century. Quarters A can be seen on the right, above the storage tanks. The bridge in the center background is the Verrazano Narrows Bridge, connecting Brooklyn and Staten Island. The skyscrapers of Brooklyn's downtown MetroTech are visible in the background. (Author's collection.)

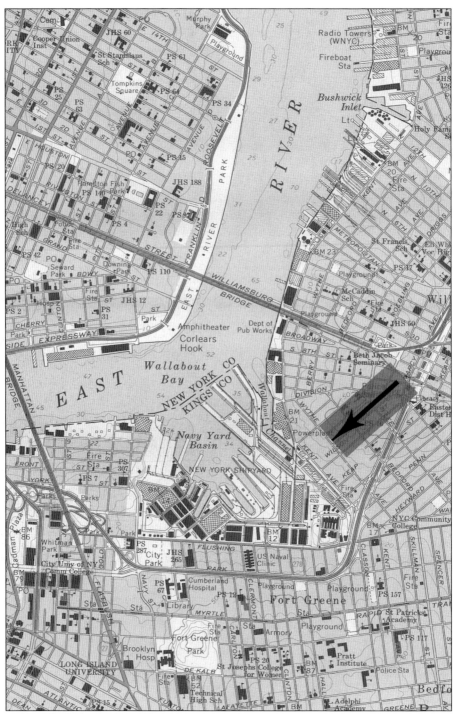

MAP OF BNY. This map shows the present-day BNY and its surroundings The map is oriented with the compass, with due north at the top of the page. Manhattan is in the upper left. Along the northern edge of the map is the place where the USS Monitor was built: Newtown Creek, which runs into the East River. (USGS.)

SATELLITE SHOT. This Soviet satellite photograph of BNY and its surroundings was taken in 1988. (www.terraserver.com.)

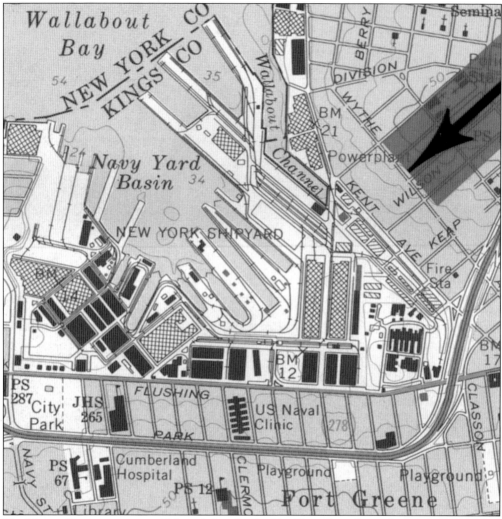

No Major Construction. Although hundreds of buildings have been demolished, there has been no major construction at the yard since the navy left. This recent map shows BNY as it appears today. (USGS.)

CIVILIAN VESSELS. The ships seen on the left in this 1988 Soviet satellite photograph are civilian vessels brought to BNY for repairs, long after the navy had ceded control. (www.terraserver.com.)

SAND STREET GATE, 1999. The Sand Street Gate serves today as the entrance to an impound lot for towed vehicles. (Author's collection.)

VACANT QUARTERS. The officers' housing along Flushing Avenue has been abandoned and is slowly decaying. (Author's collection.)

FROM 300-PLUS DOWN TO 40. This view was taken near the main gate, looking southwest. At the end of the street and out of sight, Quarters A is to the right and the Sand Street Gate is to the left. In 1999, 40 buildings containing over 4.1 million square feet of space still remained at the yard, down from more than 300 buildings in 1941. (Author's collection.)

GUARDING THE REMNANTS. A phalanx of city-owned garbage trucks guard the remnants of the yard. The building on the left, a vast open space once used for the construction of segments of warships, is abandoned. (Author's collection.)

BUSINESSES AND JOBS. This recent photograph shows a ship in the middle foreground undergoing repairs. At BNY in 1999, 175 different businesses were in operation, employing some 2,500 people. The area proposed for a movie studio is in the foreground. (Author's collection.)

STACKS AND PILES. In addition to storing vehicles, sand is stored at the yard. Ashes to ashes, dust to dust . . .. (Author's collection.)

Appendix: Ships Constructed at the Brooklyn Navy Yard

Name	Type	Begun	Launched	Commissioned
Ohio	74-gun ship of the line; 2,757 tons	1817	May 30, 1820	
Savannah	44-gun frigate	1820	May 5, 1842	
Sabine	44-gun frigate	1822	Feb 3, 1855	
Vincennes	18-gun sloop of war; 700 tons	1825	Apr 27, 1826	
Fairfield	18-gun sloop of war	1826	Jun 28, 1828	
Lexington	18-gun sloop of war; 691 tons	1825	Mar 9, 1826	
Peacock	18-gun, 2nd-class sloop of war		Sep 30, 1828	
Morris	Revenue cutter	1831	Jun 30, 1831	
Enterprise	10-gun schooner; 194 tons	1831	Oct 26, 1831	
Dolphin	10-gun brig; 224 tons	1836	Jun 17, 1836	
Pilot	2-gun schooner	1836	Sep 1836	
Fulton	9-gun paddle wheel steamer	1835	May 18, 1837	
Levant	18-gun sloop of war	1837	Dec 28, 1837	
Decatur	16-gun sloop of war	1838	Apr 9, 1839	
Missouri	10-gun paddle wheel steamer; 1,700 Tons	1839	Jan 7, 1841	
Somers	10-gun brig	1842	Apr 16, 1842	
San Jacinto	6-gun sloop of war	1847	Apr 16, 1850	
Albany	20-gun sloop of war; 1,064 tons	1843	Jun 27, 1846	
Fulton	9-gun paddle wheel steamer		Aug 30, 1851	
Niagara	40-gun steamer frigate	1854	Feb 23, 1856	
Iroquois	6-gunsloop of war	1858	Apr 12, 1859	
Oneida	9-gun sloop of war	1861	Nov 20, 1861	
Octorora	6-gun side-wheeler double-ender	1861	Dec 7, 1861	
Adirondack	9-gun steam sloop of war	1861	Feb 22, 1862	
Lackawanna	9-gun steam sloop of war	1862	Aug 9, 1862	
Ticonderoga	9-gun steam sloop of war	1861	Oct 16, 1862	
Shamrock	8-gun side-wheeler double-ender	1862	Apr 17, 1863	
Mackinaw	8-gun side-wheeler double-ender	1862	Apr 22, 1863	
Peoria	8-gun side-wheeler double-ender	1862	Oct 29, 1863	
Tullahoma	8-gun side-wheeler double-ender	1863	Nov 28, 1863	
Algonquin	12-gun side-wheeler double-ender	1863	Dec 31, 1831	
Maumee	7-gun steam sloop of war; 593 tons	1862	Jul 2, 1863	
Nyack	3-gun steam sloop of war	1862	Oct 6, 1863	
Kalamazoo	Double-turreted monitor	1863		
Ontario	Screw steam sloop	1863		
Madawaska	15-gun steam sloop of war	1863	Jul 8, 1865	
Wampanoag	Screw steam sloop	1863	Dec 15, 1864	
Quinnebaug	10-gun steam sloop of war	1864	Mar 31, 1866	
Mosholu	13-gun steam sloop of war	1864	1865	
Java	Screw steam sloop	1864		
Kenosha	Screw steamer	1867	Aug 8, 1868	
Alarm	Torpedo boat	1872	Nov 13, 1873	
Swatara	6-gun steam sloop of war		Sep 17, 1873	
Trenton	19-gun steam frigate	1875	Jan 1, 1876	Feb 14, 1877

Name	Type	Begun	Launched	Commissioned
Maine	Battleship, 2nd class 10-gun main battery steel twin screw; 6,682 tons	1888	Nov 18, 1890	Sep 17, 1895
Cincinnati	Protected cruiser, steel twin screw	1890	Nov 10, 1892	Jun 16, 1894
Terror	Double-turreted monitor		Mar 24, 1883	Apr 15, 1896
Miantonomah	Double-turreted monitor		Dec 5, 1876	Oct 27, 1891
Puritan	Double-turreted monitor		Dec 6, 1882	Dec 10, 1896
Penacook	Steel yard tug	1898	Oct 29, 1898	May 7, 1899
Connecticut	Battleship, BB 18; 16,000 tons	1903	Sep 29, 1904	Sep 29, 1906
Vestal	Fleet collier; 12,000 tons	1907	May 19, 1908	Oct 4, 1909
Florida	Battleship, BB 30; 22,000 tons	1909	May 12, 1910	Sep 15, 1911
New York	Battleship, BB 34; 27,000 tons	1911	Oct 30, 1912	Apr 15, 1914
Arizona	Battleship, BB 39; 31,400 tons	1914	Jun 19, 1915	Nov 17, 1916
New Mexico	Battleship, BB 40; 32,000 tons	1915	Apr 23, 1917	May 20, 1918
Tennessee	Battleship, BB 43; 32,000 tons	1917	Apr 30, 1919	Jun 3, 1920
South Dakota	Battleship; 43,200 tons	1920	Scrapped	
Indiana	Battleship; 43,200 tons	1920	Scrapped	
Pensacola	Heavy cruiser, CL 24; 10,000 tons	1926	Apr 25, 1929	Feb. 6, 1930
VF 221	Covered freight lighter	1932	Sep 28, 1932	Apr 10, 1933
New Orleans	Heavy cruiser, CA 32; 10,000 tons	1931	Apr 12, 1933	Feb 15, 1934
Hull	Destroyer, DD 350; 1,395 tons	1933	Jan 31, 1934	Jan 11, 1935
Dale	Destroyer, DD 353; 1,395 tons	1934	Jan 23, 1935	Jun 17, 1935
Erie	Gunboat, PG 50; 2,000 tons	1934	Jan 29, 1936	Jul 1, 1936
Brooklyn	Light cruiser, CL 40; 10,000 tons	1935	Nov 30, 1936	Sep 30, 1937
Alexander Hamilton	Coast guard cutter; 2,000 tons	1935	Jan 6, 1937	Mar 4, 1937
John C. Spencer	Coast guard cutter; 2,000 tons	1935	Jan 6, 1937	Mar 1, 1937
Honolulu	Light cruiser, CL 48; 10,000 tons	1935	Aug 26, 1937	Jun 15, 1938
Helena	Light cruiser, CL 50; 10,000 tons	1936	Aug 27, 1938	Sep 18, 1939
North Carolina	Battleship, BB 55; 35,000 tons	1937	Jun 13, 1940	Apr 9, 1941
Iowa	Battleship, BB 61; 45,000 tons	1940	Aug 27, 1942	Feb 22, 1943
Missouri	Battleship, BB 63; 45,000 tons	1941	Jan 29, 1944	Jun 11, 1944
YR 34-35	Floating workshops	1941	Nov 25, 1941	
311-314	Landing ship, tanks	1942	Dec 30, 1942	
315-318	Landing ship, tanks	1942	Jan 23, 1943	
Bennington	Aircraft carrier, CV 20; 27,100 tons	1942	Feb 26, 1944	Aug 6, 1944
Bon Homme Richard	Aircraft carrier, CV 31; 27,100 tons	1943	Apr 29, 1944	Nov 26, 1944
Franklin D. Roosevelt	Aircraft carrier, CVB 42; 45,000 tons	1943	Apr 29, 1945	Oct 27, 1945
Kearsarge	Aircraft carrier, CV 33; 27,100 tons	1944	May 5, 1945	Mar 2, 1946
Oriskany	Aircraft carrier, CV 34; 37,000 tons	1944	Oct 13, 1945	Sep 25, 1950
Saratoga	Aircraft carrier, CVA 60; 56,000 tons	1952	Oct 8, 1955	Apr 14, 1956
Independence	Aircraft carrier, CVA 62; 56,300 tons	1955	Jun 6, 1958	Jan 10, 1959
Constellation	Aircraft carrier, CVA 64; 60,000 tons	1957	Oct 8, 1960	Oct 27, 1961
Raleigh	Amphibious transport, dock LPD 1	1960	Mar 17, 1962	
Vancouver	Amphibious transport, dock LPD 2	1960	Sep 15, 1962	
La Salle	Amphibious transport, dock LPD 3	1962	Aug 3, 1963	
Austin	Amphibious transport, dock LPD 4	1963	Jun 27, 1964	Mar 1965
Ogden	Amphibious transport, dock LPD 5	1963	Jun 27, 1964	Jun 19, 1965
Duluth	Amphibious transport, dock LPD 6	1963	Aug 14, 1965	Dec 16, 1965